Mr. C and the Brainies

SOLVING SUBTRACTION

THE ACON'S WAY

PANCHITO FONTILLAS, ED.S.

Printed in the United State of America

Library of Congress Control Number: 2026906695

ISBN 978-1-7360134-7-2

Mr. C and the Brainies: Solving Subtraction the ACON's Way
Design, Illustrated, and Written by Panchito Fontillas, Ed.S.

Welcome back. Today, we will learn different ways to subtract numbers.

That is right. After reading and doing the exercises in this book, you will master the principles of subtractions and do problems involving subtraction

Suppose that the problem is 95—28. How will you solve this problem?

$$\begin{array}{r} 95 \\ -\ 28 \\ \hline ? \end{array}$$

There are different ways to solve this problem. The first way is to use the Compensation and Place Value Decomposition Technique.

$$\begin{array}{r} 95 \\ -\ 28 \\ \hline \boxed{?} \end{array}$$

First, express nine-five into its place value equivalent like this.

$$\begin{array}{r} 95 \\ -\ 28 \\ \hline \boxed{?} \end{array} \qquad \begin{array}{r} 90\ 5 \\ -\ 2\ 8 \\ \hline \boxed{} \end{array}$$

The next step is to round 28 to the nearest tens place like this.

$$\begin{array}{r} 95 \\ -\,28 \\ \hline ? \end{array} \qquad \begin{array}{rr} 90 & 5 \\ -\,30 & \\ \hline & \end{array}$$

That is right. Now, find the difference between 28 and 30. Write the difference of 28 and 30 below five like this.

$$\begin{array}{r} 95 \\ -\,28 \\ \hline ? \end{array} \qquad \begin{array}{rr} 90 & 5 \\ -\,30 & 2 \\ \hline & \end{array}$$

Now, find the difference between 90 and 30. Write the difference of 90 and 30 inside the box like this.

$$\begin{array}{r} 9\,5 \\ -\ 2\,8 \\ \hline \boxed{?} \end{array} \qquad \begin{array}{r} 90\ \ 5 \\ -\ 30\ \ 2 \\ \hline \boxed{60\ \ \ } \end{array}$$

Now, here is the twist. You are to find the sum of 5 and 2. Write the sum of 5 and 2 on the ones place like this.

$$\begin{array}{r} 9\,5 \\ -\ 2\,8 \\ \hline \boxed{?} \end{array} \qquad \begin{array}{r} 90\ \ 5 \\ -\ 30\ \ 2^{+} \\ \hline \boxed{60\ \ 7} \end{array}$$

For the last step add 60 and 7.
Replace the question mark with the
sum of 60 and 7 like this!

$$\begin{array}{r} 90\ \ 5 \\ -\ 30\ \ 2 \\ \hline \boxed{60\ \ 7} \end{array} \quad +$$

$$\begin{array}{r} 95 \\ -\ 28 \\ \hline \boxed{67} \end{array}$$

Let us have a quiz!

Here are your first four questions.

$$\begin{array}{r} 71 \\ -\,55 \\ \hline \square \end{array} \qquad \begin{array}{r} 43 \\ -\,19 \\ \hline \square \end{array}$$

$$\begin{array}{r} 57 \\ -\,38 \\ \hline \square \end{array} \qquad \begin{array}{r} 84 \\ -\,66 \\ \hline \square \end{array}$$

Here are your second set of questions.

$$\begin{array}{r} 86 \\ -\ 28 \\ \hline \square \end{array}$$

$$\begin{array}{r} 62 \\ -\ 39 \\ \hline \square \end{array}$$

$$\begin{array}{r} 73 \\ -\ 45 \\ \hline \square \end{array}$$

$$\begin{array}{r} 54 \\ -\ 27 \\ \hline \square \end{array}$$

Solve these four questions!

$$\begin{array}{r} 62 \\ -\ 58 \\ \hline \end{array}$$

$$\begin{array}{r} 86 \\ -\ 39 \\ \hline \end{array}$$

$$\begin{array}{r} 75 \\ -\ 37 \\ \hline \end{array}$$

$$\begin{array}{r} 94 \\ -\ 18 \\ \hline \end{array}$$

You are amazing!
Now, Solve these
questions.

$$\begin{array}{r} 63 \\ -\ 48 \\ \hline \square \end{array}$$

$$\begin{array}{r} 91 \\ -\ 64 \\ \hline \square \end{array}$$

$$\begin{array}{r} 53 \\ -\ 29 \\ \hline \square \end{array}$$

$$\begin{array}{r} 72 \\ -\ 55 \\ \hline \square \end{array}$$

Wow! Keep going!
You are doing great!

$$\begin{array}{r} 76 \\ -\,57 \\ \hline \end{array}$$

$$\begin{array}{r} 82 \\ -\,35 \\ \hline \end{array}$$

$$\begin{array}{r} 63 \\ -\,47 \\ \hline \end{array}$$

$$\begin{array}{r} 94 \\ -\,58 \\ \hline \end{array}$$

$$\begin{array}{r} 77 \\ -\ 48 \\ \hline \boxed{} \end{array} \qquad \begin{array}{r} 33 \\ -\ 27 \\ \hline \boxed{} \end{array}$$

$$\begin{array}{r} 82 \\ -\ 56 \\ \hline \boxed{} \end{array}$$

Now, let us do the last set of questions.

$$\begin{array}{r} 81 \\ -\ 37 \\ \hline \boxed{} \end{array}$$

That is amazing! Now, we will learn another technique in subtracting numbers.

It is easy to learn and to follow. The technique is called as Compensation or Constant Difference Technique.

Suppose that the problem is ninety—five minus twenty –eight.

$$\begin{array}{r} 95 \\ -\ 28 \\ \hline \square \end{array}$$

Make the 8 on the ones place a ten.

Add the number that makes the 8 a ten to five, like this.

$$\begin{array}{r} 95^{+2} \\ -\ 28 \\ \hline \square \end{array}$$

Encircle the 5 and 2 together and place the sum of 5 and 2 on the ones place under 8.

$$\begin{array}{r} 9\ (5+2) \\ -\ 2\ 8 \\ \hline \boxed{\ \ 7} \end{array}$$

Round up 28 and subtract. Write the difference on the tens place.

$$\begin{array}{r} 9\ (5+2) \\ -\ 3\ 0 \\ \hline \boxed{6\ 7} \end{array}$$

You truly are magnificent!

You quickly learn two strategies on how to subtract numbers using the ACON's way.

Now, let us have a short quiz to find out if you can apply the second technique.

Now, find the difference of each expression using the second technique.

$$\begin{array}{r} 86 \\ -\ 49 \\ \hline \end{array}$$

$$\begin{array}{r} 91 \\ -\ 53 \\ \hline \end{array}$$

$$\begin{array}{r} 74 \\ -\ 35 \\ \hline \end{array}$$

$$\begin{array}{r} 83 \\ -\ 67 \\ \hline \end{array}$$

Find the difference of each problem.

$$\begin{array}{r} 61 \\ -\ 49 \\ \hline \end{array}$$

$$\begin{array}{r} 51 \\ -\ 33 \\ \hline \end{array}$$

$$\begin{array}{r} 73 \\ -\ 24 \\ \hline \end{array}$$

$$\begin{array}{r} 54 \\ -\ 39 \\ \hline \end{array}$$

Let us work on this problems as well.

$$\begin{array}{r} 41 \\ -\ 13 \\ \hline \square \end{array}$$

$$\begin{array}{r} 92 \\ -\ 79 \\ \hline \square \end{array}$$

$$\begin{array}{r} 81 \\ -\ 27 \\ \hline \square \end{array}$$

$$\begin{array}{r} 76 \\ -\ 39 \\ \hline \square \end{array}$$

$$\begin{array}{r} 68 \\ -\ 19 \\ \hline \square \end{array} \qquad \begin{array}{r} 54 \\ -\ 29 \\ \hline \square \end{array}$$

Apply your strategies to find the answers to these queries.

$$\begin{array}{r} 74 \\ -\ 25 \\ \hline \square \end{array} \qquad \begin{array}{r} 46 \\ -\ 17 \\ \hline \square \end{array}$$

$$\begin{array}{r} 42 \\ -\ 19 \\ \hline \square \end{array}$$

$$\begin{array}{r} 75 \\ -\ 48 \\ \hline \square \end{array}$$

Let us finish this quiz with these questions.

$$\begin{array}{r} 52 \\ -\ 29 \\ \hline \square \end{array}$$

$$\begin{array}{r} 60 \\ -\ 48 \\ \hline \square \end{array}$$

And now, we will learn the third strategy on how to subtract numbers.

This strategy is very easy. This strategy is called as Partial Negative Technique.

Suppose that the problem is 95—28.

$$\begin{array}{r} 95 \\ -\ 28 \\ \hline \boxed{?} \end{array}$$

Rewrite the problem to expanded form like this.

$$\begin{array}{r} 95 \\ -\ 28 \\ \hline \boxed{?} \end{array}$$

$$\begin{array}{rr} 90 & 5 \\ -\ 20 & 8 \\ \hline \end{array}$$

$\boxed{}$

Subtract 20 from 90
and write the difference inside
the box on the Tens Place.

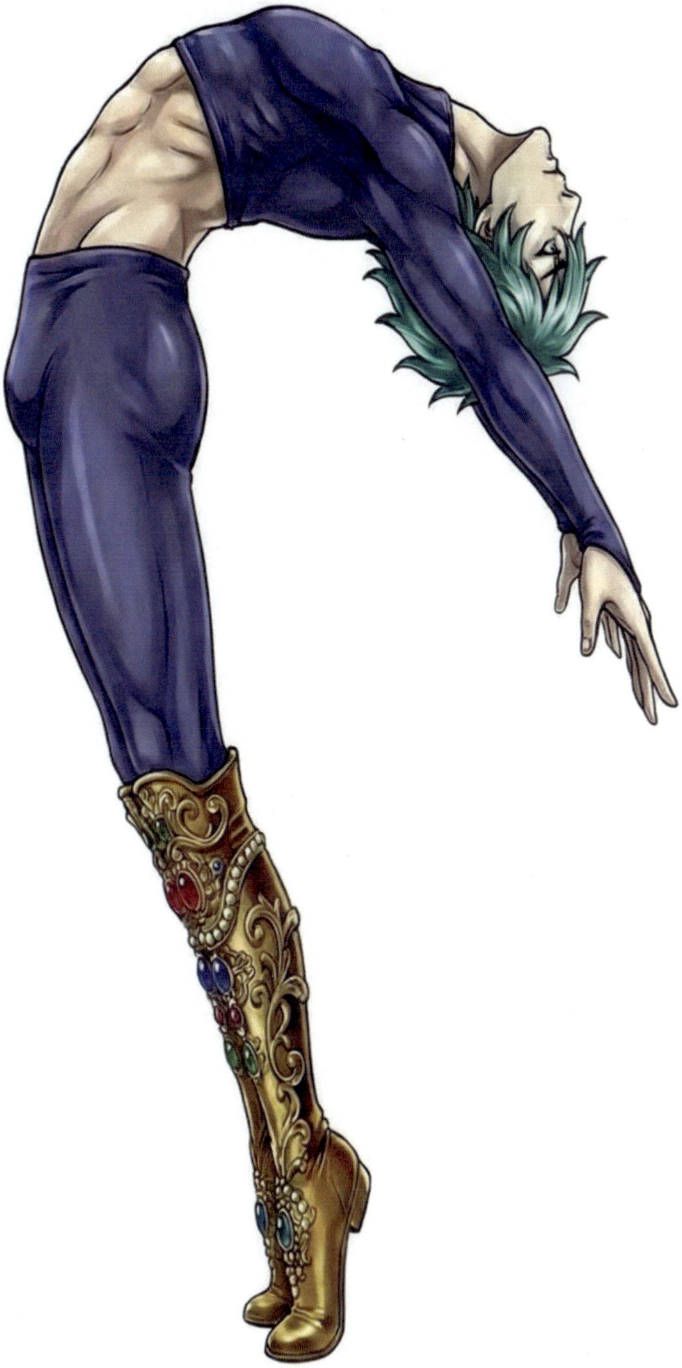

$$\begin{array}{r} 90\ 5 \\ -\ 20\ 8 \\ \hline \boxed{70} \end{array}$$

Subtract eight from
five using the number line.

	90	5
−	20	8
	70	

Write the difference
of 5 and 8 like this.

```
  90  5
− 20  8
-------
  70 −3
```

Decompose 70 by 60 and 10 like this

$$\begin{array}{r} 90\ \ 5 \\ -\ 20\ \ 8 \\ \hline 70\ -3 \end{array}$$

$$70 \rightarrow 60 \quad 10$$

Subtract –3 from 10 like this and write the difference below.

$$\begin{array}{r} 90\ 5 \\ -\ 20\ 8 \\ \hline 70\ \ \ \ \end{array}$$

60 10 - 3

7

Now, add 60 and 7. Therefore, the answer to the problem is 67.

$$\begin{array}{r} 90\;\;5 \\ -\;20\;\;8 \\ \hline 70 \end{array}$$

70 → 60 and 10 - 3

10 - 3 = 7

$$\begin{array}{r} 95 \\ -\;28 \\ \hline \boxed{67} \end{array}$$

Now that you have completed the exercises using the Partial Negative technique, we will have a short quiz.

Find the difference of each expression using the Partial Negative Technique.

$$\begin{array}{r} 62 \\ -\ 39 \\ \hline \end{array}$$

$$\begin{array}{r} 57 \\ -\ 28 \\ \hline \end{array}$$

$$\begin{array}{r} 84 \\ -\ 66 \\ \hline \end{array}$$

$$\begin{array}{r} 84 \\ -\ 29 \\ \hline \end{array}$$

Let us try these solving these problems.

$$\begin{array}{r} 71 \\ -\ 19 \\ \hline \square \end{array} \qquad \begin{array}{r} 82 \\ -\ 67 \\ \hline \square \end{array}$$

$$\begin{array}{r} 51 \\ -\ 29 \\ \hline \square \end{array} \qquad \begin{array}{r} 66 \\ -\ 37 \\ \hline \square \end{array}$$

Find the difference of each expression using the Partial Negative Technique.

$$\begin{array}{r} 77 \\ -\ 48 \\ \hline \end{array}$$

$$\begin{array}{r} 93 \\ -\ 67 \\ \hline \end{array}$$

$$\begin{array}{r} 95 \\ -\ 39 \\ \hline \end{array}$$

$$\begin{array}{r} 61 \\ -\ 18 \\ \hline \end{array}$$

Let us try move on to these problems.

$$\begin{array}{r} 63 \\ -\,55 \\ \hline \square \end{array}$$

$$\begin{array}{r} 93 \\ -\,77 \\ \hline \square \end{array}$$

$$\begin{array}{r} 80 \\ -\,44 \\ \hline \square \end{array}$$

$$\begin{array}{r} 81 \\ -\,38 \\ \hline \square \end{array}$$

$$\begin{array}{r} 85 \\ -\ 28 \\ \hline \end{array}$$

$$\begin{array}{r} 62 \\ -\ 27 \\ \hline \end{array}$$

$$\begin{array}{r} 81 \\ -\ 39 \\ \hline \end{array}$$

Let us find the answer to each problem to conclude this quiz.

$$\begin{array}{r} 71 \\ -\ 39 \\ \hline \end{array}$$

Congratulations! Now, we will proceed to the next level.

We will learn how to subtract three-digits numbers.

That is right. We will
use the same strategies.

Our first problem is this!

$$\begin{array}{r} 864 \\ -\ 478 \\ \hline \boxed{?} \end{array}$$

This looks so challenging.

$$\begin{array}{r} 864 \\ -\,478 \\ \hline \boxed{?} \end{array}$$

Yeah, but this is actually easy.

$$\begin{array}{r} 864 \\ -\ 478 \\ \hline ? \end{array}$$

Round up 78 to the tens place like this.

$$\begin{array}{r} 864 \\ -\ 478 \\ \hline ? \end{array} \qquad \begin{array}{r} 864 \\ -\ 480 \\ \hline ? \end{array}$$

Add the difference of
80 and 78 to 64 like this.

$$\begin{array}{r} 864 \\ -\,478 \\ \hline ? \end{array}$$

$$\begin{array}{r} 866 \\ -\,480 \\ \hline ? \end{array}$$

Subtract 0 from 6 and
write the difference like this.

$$\begin{array}{r} 866 \\ -\ 480 \\ \hline \boxed{\ ?\ } \end{array}$$

$$\begin{array}{r} 866 \\ -\ 480 \\ \hline \boxed{\quad 6} \end{array}$$

Round 48 up like this.

Subtract 48 from 50
and add the difference to 86
like this.

$$\begin{array}{r} 866 \\ -\ 480 \\ \hline \boxed{\quad 6} \end{array}$$

$$\begin{array}{r} 886 \\ -\ 500 \\ \hline \boxed{\quad 6} \end{array}$$

But what if I were to subtract from left to right using the same technique. Will it work?.

$$\begin{array}{r} 886 \\ -\ 500 \\ \hline \boxed{386} \end{array}$$

But what if I were to subtract from left to right using the same technique. Will it work?.

$$\begin{array}{r} 864 \\ -\ 478 \\ \hline \boxed{} \end{array}$$

Of course it will work. We can solve this problem using mental math technique called Constance Difference.

$$\begin{array}{r} 864 \\ -\ 478 \\ \hline \square \end{array}$$

First, subtract 4 from and rewrite the expression like this.

$$\begin{array}{r} 864 \\ -\ 478 \\ \hline \boxed{} \end{array}$$

$$\begin{array}{r} 464 \\ -\ 78 \\ \hline \boxed{} \end{array}$$

Round 78 to 80. Add the difference of 78 and 80 to 464 like this.

$$\begin{array}{r} 864 \\ -\ 478 \\ \hline \square \end{array}$$

$$\begin{array}{r} 464 \\ -\ 78 \\ \hline \square \end{array}$$

$$\begin{array}{r} 464 \\ -\ 78 \\ \hline \square \end{array} \qquad \begin{array}{r} 466 \\ -\ 80 \\ \hline \square \end{array}$$

Balance the expression by adding 20 to 80 like this. Do not forget to add the 20 to 466.

$$\begin{array}{r} 864 \\ -\,478 \\ \hline \square \end{array}$$

$$\begin{array}{r} 464 \\ -\,78 \\ \hline \square \end{array}$$

$$\begin{array}{r} 464 \\ -\,78 \\ \hline \square \end{array}$$

$$\begin{array}{r} 466 \\ -\,80 \\ \hline \square \end{array}$$

$$\begin{array}{r} 466 \\ -\,80 \\ \hline \square \end{array}$$

$$\begin{array}{r} 486 \\ -\,100 \\ \hline \square \end{array}$$

Finally, subtract 100
from 486 like this.

$$\begin{array}{r} 864 \\ -\ 478 \\ \hline \square \end{array} \qquad \begin{array}{r} 464 \\ -\ 78 \\ \hline \square \end{array} \qquad \begin{array}{r} 464 \\ -\ 78 \\ \hline \square \end{array}$$

$$\begin{array}{r} 466 \\ -\ 80 \\ \hline \square \end{array} \qquad \begin{array}{r} 466 \\ -\ 80 \\ \hline \square \end{array} \qquad \begin{array}{r} 486 \\ -\ 100 \\ \hline 386 \end{array}$$

Purely amazing!

Now, it is time for us to have a quiz.

Now, let us find solutions to these problems.

$$\begin{array}{r} 773 \\ -\ 487 \\ \hline \square \end{array}$$

$$\begin{array}{r} 932 \\ -\ 675 \\ \hline \square \end{array}$$

$$\begin{array}{r} 954 \\ -\ 396 \\ \hline \square \end{array}$$

$$\begin{array}{r} 613 \\ -\ 185 \\ \hline \square \end{array}$$

$$\begin{array}{r} 924 \\ -\ 386 \\ \hline \end{array}$$

Solve this too.

$$\begin{array}{r} 625 \\ -\ 189 \\ \hline \end{array}$$

$$\begin{array}{r} 712 \\ -\ 256 \\ \hline \end{array}$$

$$\begin{array}{r} 632 \\ -\ 499 \\ \hline \end{array}$$

Solve these as well.

$$\begin{array}{r} 823 \\ -\ 647 \\ \hline \end{array}$$

$$\begin{array}{r} 922 \\ -\ 285 \\ \hline \end{array}$$

$$\begin{array}{r} 921 \\ -\ 678 \\ \hline \end{array}$$

$$\begin{array}{r} 526 \\ -\ 278 \\ \hline \end{array}$$

These problems too.

$$\begin{array}{r} 642 \\ -\ 289 \\ \hline \square \end{array}$$

$$\begin{array}{r} 821 \\ -\ 148 \\ \hline \square \end{array}$$

$$\begin{array}{r} 426 \\ -\ 187 \\ \hline \square \end{array}$$

$$\begin{array}{r} 826 \\ -\ 289 \\ \hline \square \end{array}$$

And of course, these questions.

$$\begin{array}{r} 723 \\ -\ 197 \\ \hline \end{array}$$

$$\begin{array}{r} 725 \\ -\ 359 \\ \hline \end{array}$$

$$\begin{array}{r} 726 \\ -\ 357 \\ \hline \end{array}$$

$$\begin{array}{r} 723 \\ -\ 269 \\ \hline \end{array}$$

You are excellent!

Now, we will explore the other techniques in solving subtraction problems with three digits numbers.

We will use the Partial
Difference Technique.

Great! Now, suppose
our problem is 926—489.

How will you solve this?

$$\begin{array}{r} 926 \\ -\ 489 \\ \hline \boxed{} \end{array}$$

Well, I can help you with that.

First, rewrite
your numbers into
expanded form.

$$\begin{array}{r} 926 \\ -\ 489 \\ \hline \end{array}$$

$$\begin{array}{rrr} 900 & 20 & 6 \\ -\ 400 & 80 & 9 \\ \hline \end{array}$$

Mentally, make 9 a 10. Add the difference of 9 and ten to 6.

$$\begin{array}{r} 926 \\ -\ 489 \\ \hline \square \end{array}$$

$$\begin{array}{rrr} 900 & 20 & 6 \\ -\ 400 & 80 & 9 \\ \hline \square \end{array}$$

$$\begin{array}{rrr} 900 & 20 & 7 \\ -\ 400 & 80 & \\ \hline \square \end{array}$$

Change 80 to 90.

	9	2	6
−	4	8	9

	900	20	6
−	400	80	9

	900	20	7
−	400	90	

Mentally subtract 90 from 100, Ad d its difference to 20.

$$\begin{array}{r} 926 \\ -\ 489 \\ \hline \end{array}$$

$$\begin{array}{rrr} 900 & 20 & 6 \\ -\ 400 & 80 & 9 \\ \hline \end{array}$$

$$\begin{array}{rrr} 900 & 20 & 7 \\ -\ 400 & 90 & \\ \hline \end{array}$$

$$\begin{array}{rrr} 900 & 30 & 7 \\ -\ 400 & & \\ \hline \end{array}$$

Finally, change 400 to 500 and subtract!

$$\begin{array}{r} 926 \\ -\ 489 \\ \hline \boxed{} \end{array}$$

$$\begin{array}{rrr} 900 & 20 & 6 \\ -\ 400 & 80 & 9 \\ \hline \end{array}$$

$\boxed{}$

$$\begin{array}{rrr} 900 & 20 & 7 \\ -\ 400 & 90 & \\ \hline \end{array}$$

$\boxed{}$

$$\begin{array}{rrr} 900 & 30 & 7 \\ -\ 500 & & \\ \hline \end{array}$$

$\boxed{400\ 30\ 7}$

Will we be able to use this strategy if we were to solve the expression from right to left?

Well, of course, yes!

Could you show it to us?

Now, let me show
you why the strategy works.

Suppose that the problem is 425—138.

$$\begin{array}{r} 425 \\ -\,138 \\ \hline \boxed{} \end{array}$$

$$\begin{array}{r} 425 \\ -\ 138 \\ \hline \end{array}$$

$$\begin{array}{r} 400\ 20\ 5 \\ -\ 100\ 30\ 8 \\ \hline \end{array}$$

Rewrite the expression into an expanded form.

Change 100 to 200 hundred and subtract the 200 from 400. Write the difference below.

$$\begin{array}{r} 425 \\ -\ 138 \\ \hline \end{array}$$

$$\begin{array}{r} 400\ 20\ 5 \\ -\ 100\ 30\ 8 \\ \hline \end{array}$$

$$\begin{array}{r} 400\ 20\ 5 \\ -\ 200\ 30\ 8 \\ \hline \end{array}$$

200

Change 30 to 40.

$$\begin{array}{r} 4\,2\,5 \\ -\ 1\,3\,8 \\ \hline \end{array}$$

$$\begin{array}{r} 400\ 20\ 5 \\ -\ 100\ 30\ 8 \\ \hline \end{array}$$

$$\begin{array}{r} 400\ 20\ 5 \\ -\ 200\ 40\ 8 \\ \hline 200\ \end{array}$$

$$\begin{array}{r} 4\ 2\ 5 \\ -\ 1\ 3\ 8 \\ \hline \boxed{} \end{array} \qquad \begin{array}{r} 400\ 20\ 5 \\ -\ 100\ 30\ 8 \\ \hline \boxed{} \end{array}$$

$$\begin{array}{r} 400\ 20\ 5 \\ -\ 200\ 40\ 8 \\ \hline \boxed{200} \end{array} \qquad \begin{array}{r} 400\ 80\ 5 \\ -\ 200\ 8 \\ \hline \boxed{200} \end{array}$$

Mentally subtract 40 from 100. Add the difference of 100 and 40 to 20 like this.

Mentally subtract 8 from 10. Add the difference between 10 and 8 to 5 like this.

$$\begin{array}{r} 425 \\ -\ 138 \\ \hline \square \end{array}$$

$$\begin{array}{r} 400\ 20\ 5 \\ -\ 100\ 30\ 8 \\ \hline \square \end{array}$$

$$\begin{array}{r} 400\ 20\ 5 \\ -\ 200\ 40\ 8 \\ \hline 200 \end{array}$$

$$\begin{array}{r} 400\ 80\ 5 \\ -\ 200\ \ \ \ \ 8 \\ \hline 200 \end{array} \qquad \begin{array}{l} 400\ 80\ 7 \\ -\ 200 \\ \hline 200 \end{array}$$

Finally, subtract everything.

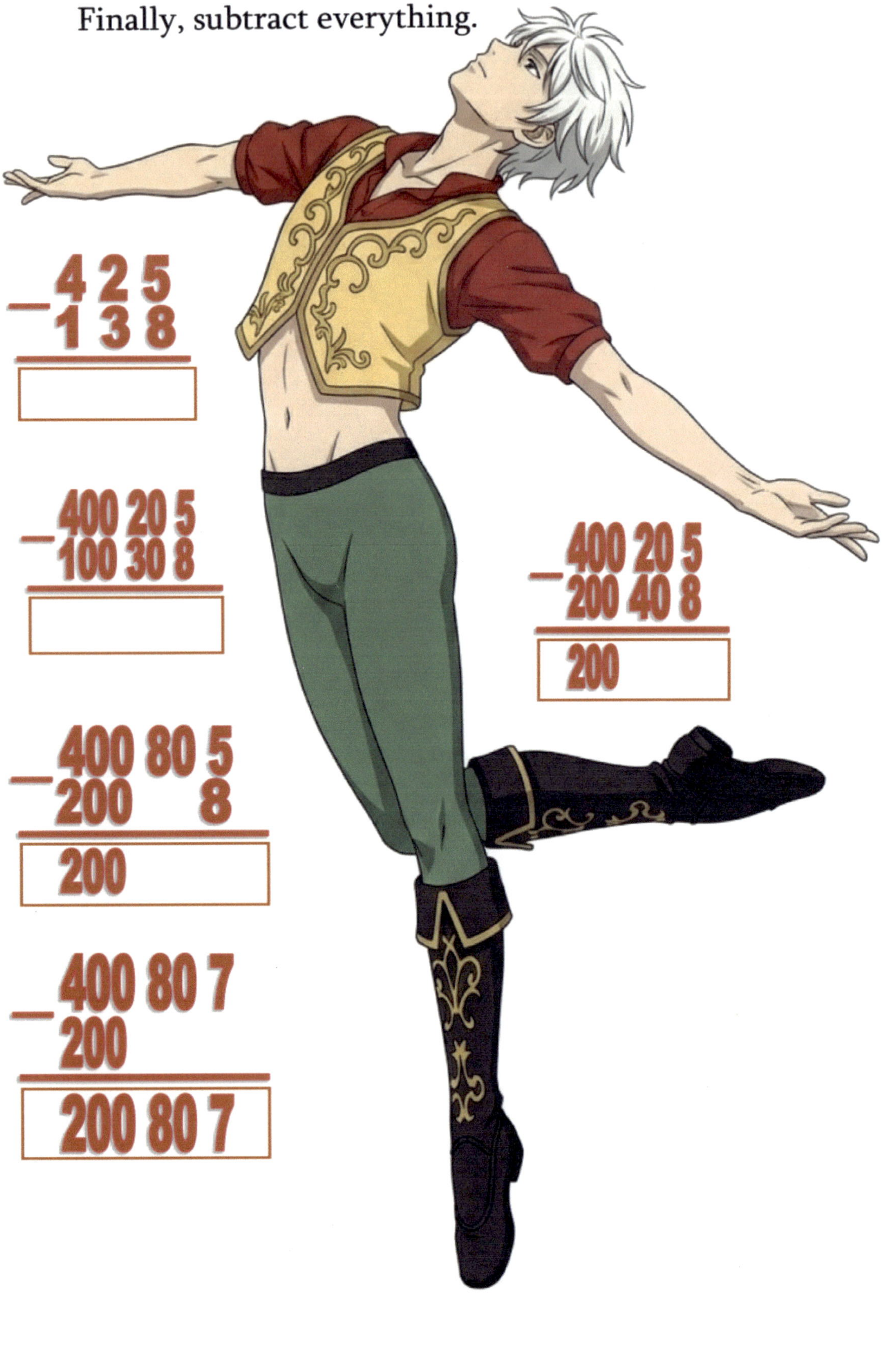

$$\begin{array}{r} 425 \\ -\ 138 \\ \hline \square \end{array}$$

$$\begin{array}{r} 400\ 20\ 5 \\ -\ 100\ 30\ 8 \\ \hline \square \end{array}$$

$$\begin{array}{r} 400\ 20\ 5 \\ -\ 200\ 40\ 8 \\ \hline 200 \end{array}$$

$$\begin{array}{r} 400\ 80\ 5 \\ -\ 200\ \quad\ 8 \\ \hline 200 \end{array}$$

$$\begin{array}{r} 400\ 80\ 7 \\ -\ 200 \\ \hline 200\ 80\ 7 \end{array}$$

Wow, this strategy is brilliant!

Let’s have a quiz.

Answer these.

$$\begin{array}{r} 821 \\ -\ 653 \\ \hline \square \end{array}$$

$$\begin{array}{r} 674 \\ -\ 185 \\ \hline \square \end{array}$$

$$\begin{array}{r} 732 \\ -\ 459 \\ \hline \square \end{array}$$

$$\begin{array}{r} 824 \\ -\ 435 \\ \hline \square \end{array}$$

Find the answers to these questions.

$$\begin{array}{r} 525 \\ -\ 296 \\ \hline \end{array}$$

$$\begin{array}{r} 962 \\ -\ 673 \\ \hline \end{array}$$

$$\begin{array}{r} 721 \\ -\ 157 \\ \hline \end{array}$$

$$\begin{array}{r} 627 \\ -\ 239 \\ \hline \end{array}$$

Solve these problems.

$$\begin{array}{r} 627 \\ -\ 239 \\ \hline \end{array}$$

$$\begin{array}{r} 962 \\ -\ 673 \\ \hline \end{array}$$

$$\begin{array}{r} 721 \\ -\ 157 \\ \hline \end{array}$$

$$\begin{array}{r} 525 \\ -\ 296 \\ \hline \end{array}$$

Find the difference.

$$\begin{array}{r} 422 \\ -\ 175 \\ \hline \square \end{array}$$

$$\begin{array}{r} 824 \\ -\ 278 \\ \hline \square \end{array}$$

$$\begin{array}{r} 661 \\ -\ 273 \\ \hline \square \end{array}$$

$$\begin{array}{r} 824 \\ -\ 376 \\ \hline \square \end{array}$$

Solve these expressions.

$$\begin{array}{r} 662 \\ -\,485 \\ \hline \square \end{array}$$

$$\begin{array}{r} 725 \\ -\,156 \\ \hline \square \end{array}$$

$$\begin{array}{r} 651 \\ -\,487 \\ \hline \square \end{array}$$

$$\begin{array}{r} 720 \\ -\,169 \\ \hline \square \end{array}$$

Are you ready to do the next challenge?

It's not my turn yet.
It's Annabelle's.

I need to hurry with this arrangement.

I wonder if she is ready....

Do I have to call her?

Hi, I am here to help
you use the third strategy.

We will learn how to use the
Partial Negative Technique in
subtracting three-digits numbers.

Yeah, what if the problem is 704—238?

$$\begin{array}{r} 704 \\ -\,238 \\ \hline \boxed{} \end{array}$$

How will we solve this using the Partial Negative Technique?

First, rewrite the numbers into expanded form.

$$\begin{array}{r} 704 \\ -\ 238 \\ \hline \end{array}$$

$$\begin{array}{r} 700\ 00\ 4 \\ -\ 200\ 30\ 8 \\ \hline \end{array}$$

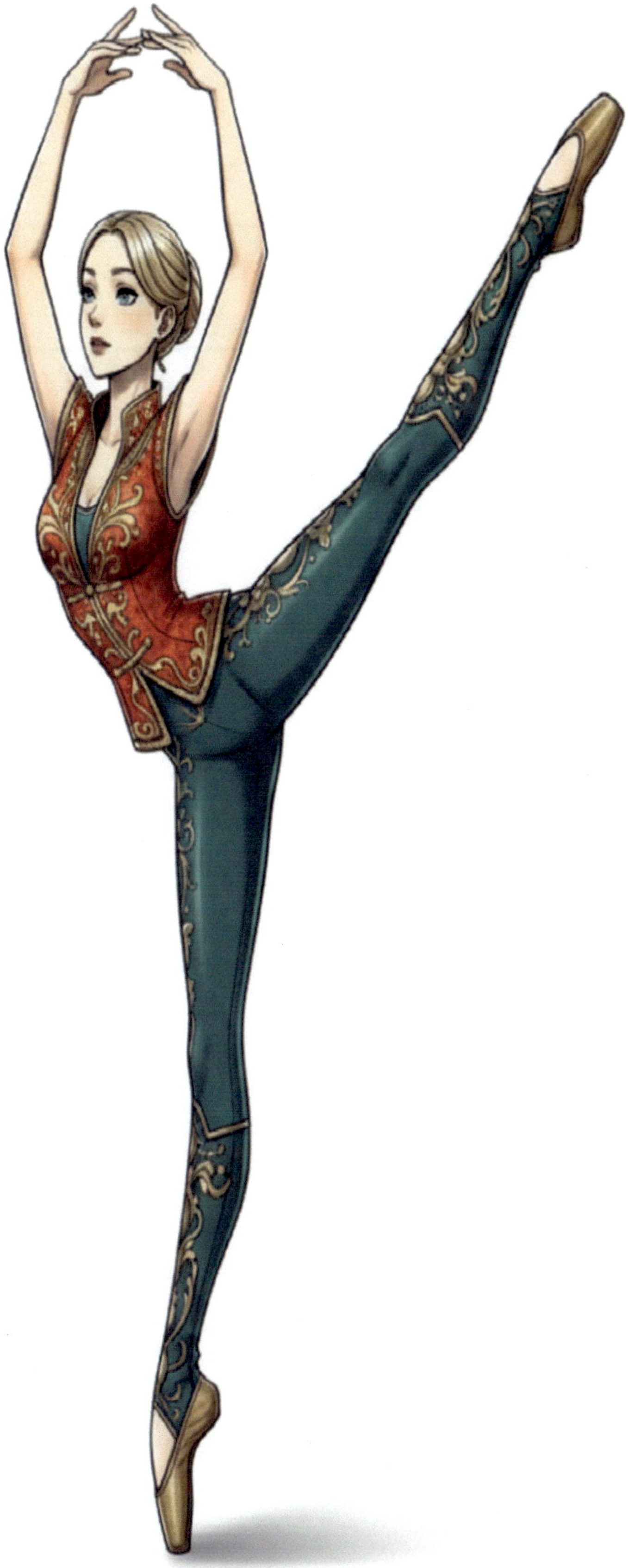

Mentally subtract 8 from 4 using the number line. Write the difference of 4 and 8 like this.

$$\begin{array}{r} 7\ 0\ 4 \\ -\ 2\ 3\ 8 \\ \hline \boxed{} \end{array}$$

$$\begin{array}{r} 700\ 00\ 4 \\ -\ 200\ 30\ 8 \\ \hline \end{array}$$

$$\begin{array}{r} 700\ \ 00\ \ 4 \\ -\ 200\ \ 30\ \ 8 \\ \hline -4 \end{array}$$

Mentally subtract 30 from 0.
Write the difference like this.

$$\begin{array}{r} 704 \\ -\ 238 \\ \hline \square \end{array}$$

$$\begin{array}{rrr} 700 & 00 & 4 \\ -\ 200 & 30 & 8 \\ \hline \end{array}$$

$$\begin{array}{rrr} 700 & 00 & 4 \\ -\ 200 & 30 & 8 \\ \hline & & -4 \end{array} \qquad \begin{array}{rrr} 700 & 00 & 4 \\ -\ 200 & 30 & 8 \\ \hline & -30 & -4 \end{array}$$

Mentally subtract 200 from 700.
Write the difference like this.

$$\begin{array}{r} 704 \\ -\ 238 \\ \hline \boxed{} \end{array}$$

$$\begin{array}{r} 700\ 00\ 4 \\ -\ 200\ 30\ 8 \\ \hline \end{array}$$

$$\begin{array}{r} 700\ \ 00\ \ 4 \\ -\ 200\ \ 30\ \ 8 \\ \hline -4 \end{array} \qquad \begin{array}{r} 700\ \ 00\ \ 4 \\ -\ 200\ \ 30\ \ 8 \\ \hline 500-30-4 \end{array}$$

Decompose 500 like this.

```
  704
- 238
 ____
 [  ]
```

```
  700 00 4
- 200 30 8
 _________
```

```
  700 00 4
- 200 30 8
 _________
         -4
```

```
  700 00 4
- 200 30 8
 _________
 500-30-4
  /    \
400   100
```

Subtract 30 from 100.

$$\begin{array}{r} 704 \\ -\ 238 \\ \hline \square \end{array}$$

$$\begin{array}{r} 700\ 00\ 4 \\ -\ 200\ 30\ 8 \\ \hline \end{array}$$

$$\begin{array}{r} 700\ \ 00\ \ 4 \\ -\ 200\ \ 30\ \ 8 \\ \hline -4 \end{array}$$

$$\begin{array}{r} 700\ \ 00\ \ 4 \\ -\ 200\ \ 30\ \ 8 \\ \hline 500-30-4 \end{array}$$

500 → 400 and 100

400 100 -30

↓

70

Decompose 70 like this.

$$
\begin{array}{r}
700\ \ 00\ \ 4 \\
-\ 200\ \ 30\ \ 8 \\
\hline
500-30-4
\end{array}
$$

500 → 400, 100

400 100 -30

100 -30 → 70

70 → 60, 10

60 10

Subtract 4 from 10.

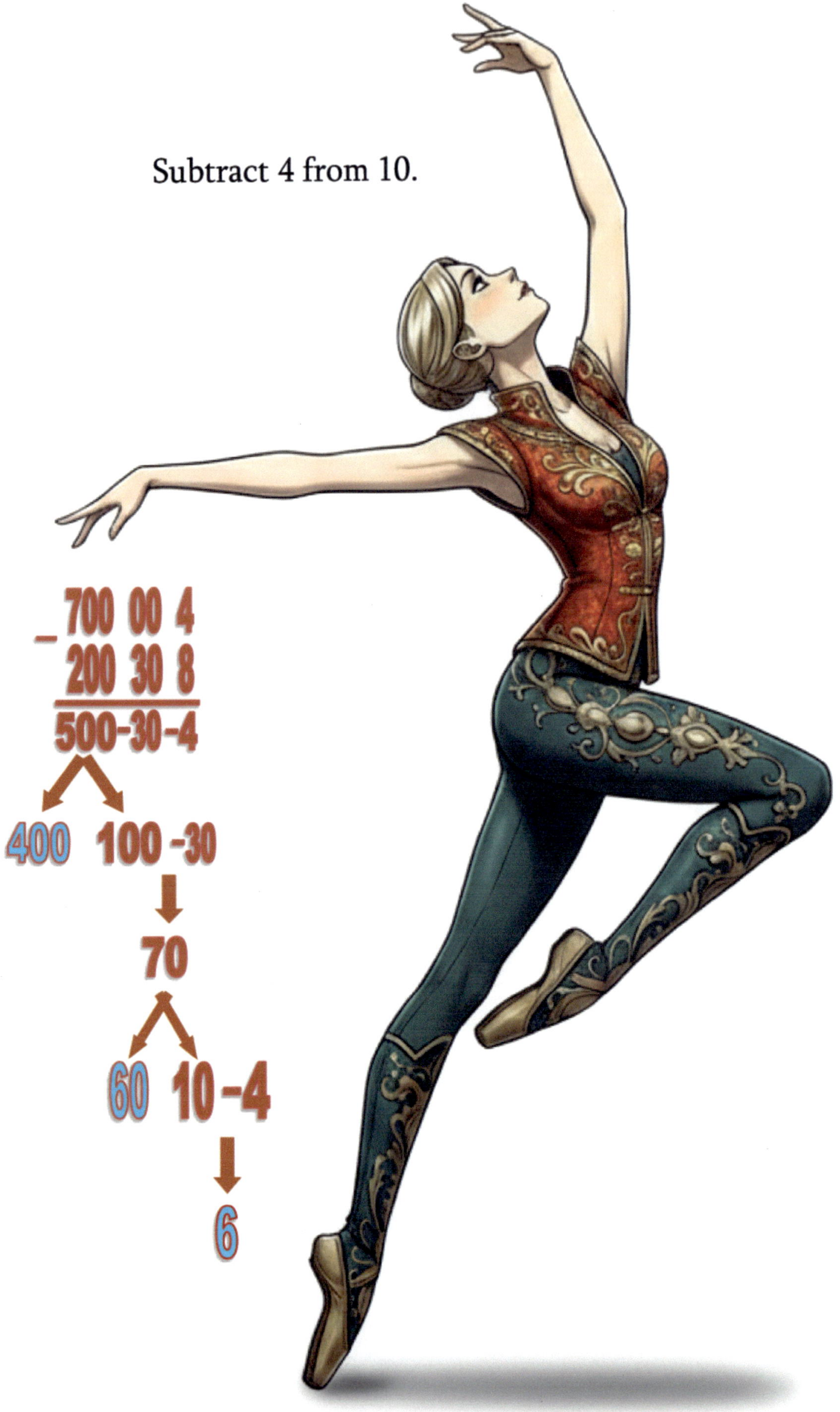

Therefore, the answer
to 704—208 is 466.

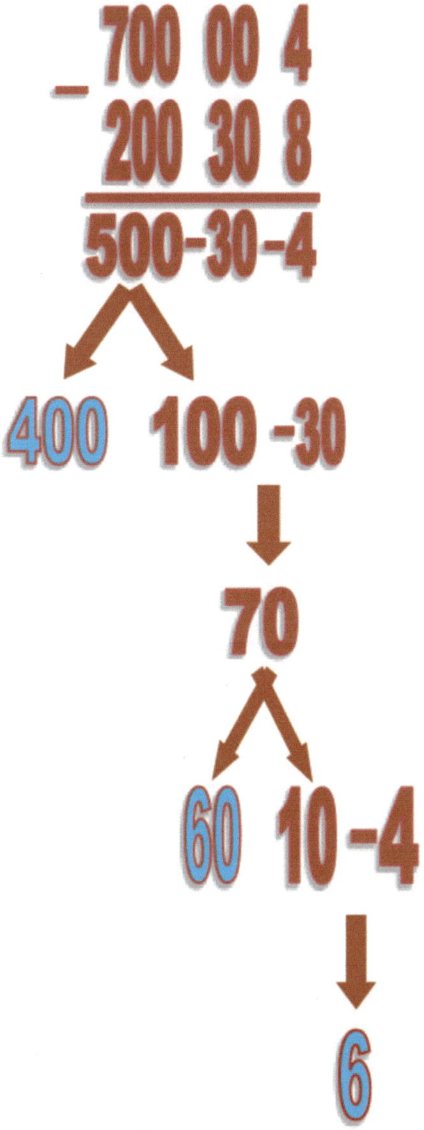

Well, it looks like that Annabelle is done.
I guess it is now my turn. I wonder what I'll be asked to explain.

Well, we are almost done.

Now, we will find out if Partial Negative Technique would work if we were to solve numbers from left to right.

And I will be here to assist you with that.

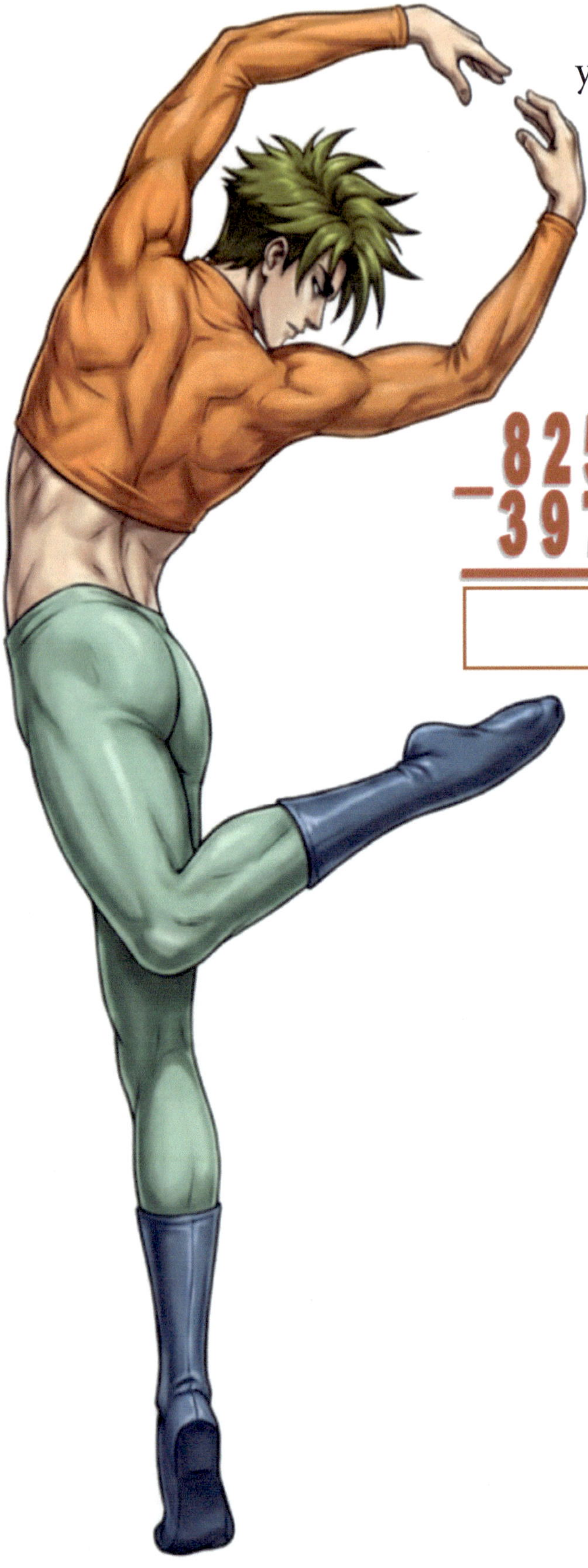

$$\begin{array}{r} 825 \\ -\,397 \\ \hline \square \end{array}$$

Express the numbers in expanded form.

$$\begin{array}{r} 825 \\ -\,397 \\ \hline \end{array}$$

$$\begin{array}{rrr} 800 & 20 & 5 \\ -\,300 & 90 & 7 \\ \hline \end{array}$$

Mentally subtract 300 from 800 and write the difference inside the rectangle.

$$\begin{array}{r} 825 \\ -\,397 \\ \hline \boxed{} \end{array}$$

$$\begin{array}{rrr} 800 & 20 & 5 \\ -\,300 & 90 & 7 \\ \hline \boxed{500} & & \end{array}$$

Mentally subtract 90 from 20 and write its difference.

$$
\begin{array}{r}
825 \\
-\ 397 \\
\hline
\boxed{}
\end{array}
$$

$$
\begin{array}{rrr}
800 & 20 & 5 \\
-\ 300 & 90 & 7 \\
\hline
500 & -70 &
\end{array}
$$

Mentally subtract 7 from 5
and write its difference.

$$\begin{array}{r} 825 \\ -\,397 \\ \hline \boxed{} \end{array}$$

$$\begin{array}{rrr} 800 & 20 & 5 \\ -\,300 & 90 & 7 \\ \hline 500 & -70 & -2 \end{array}$$

Decompose 500 like this.

Subtract 70 from 100 and write its difference.

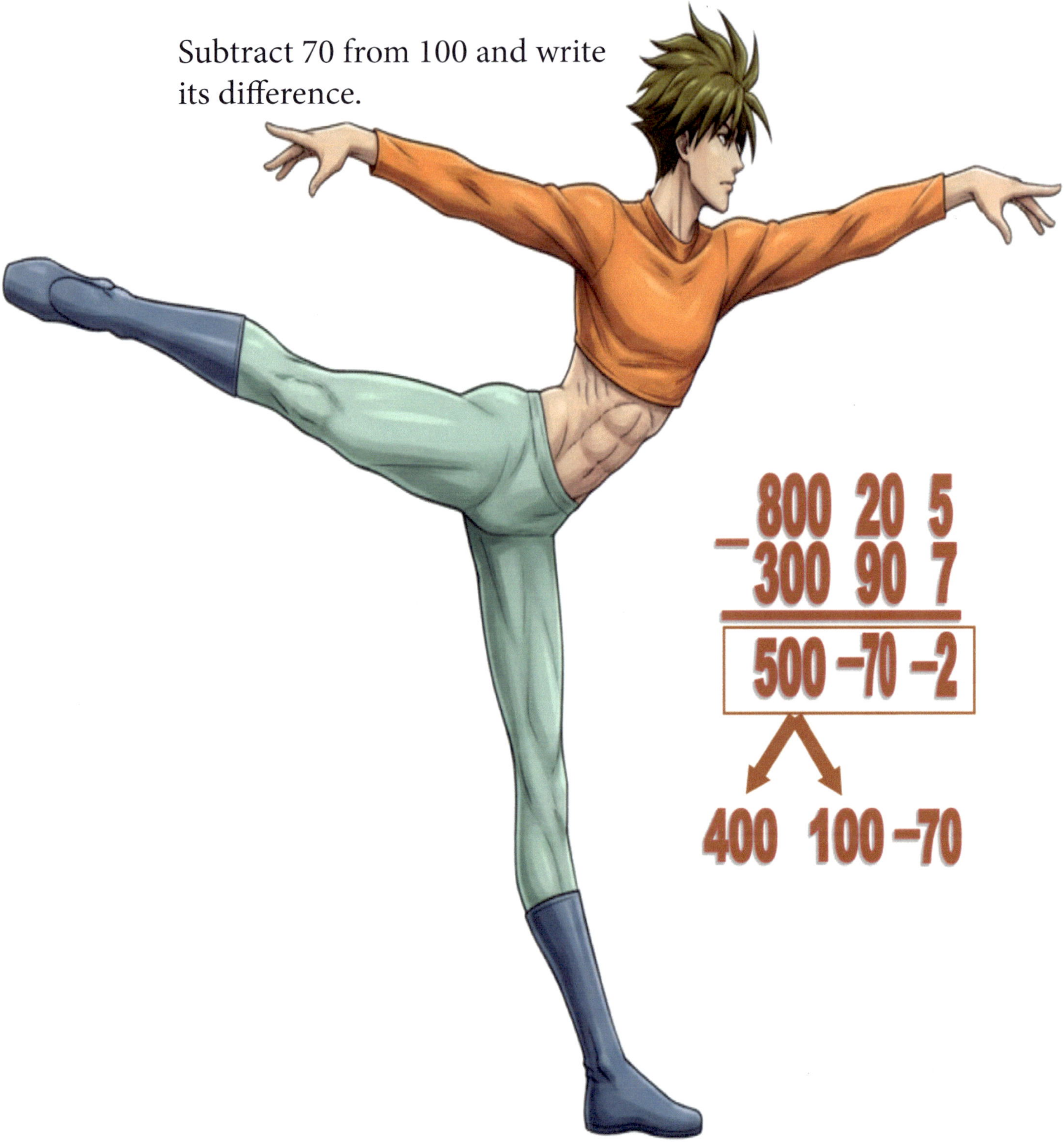

Subtract 70 from 100 and
write its difference.

Decompose 30 like this.

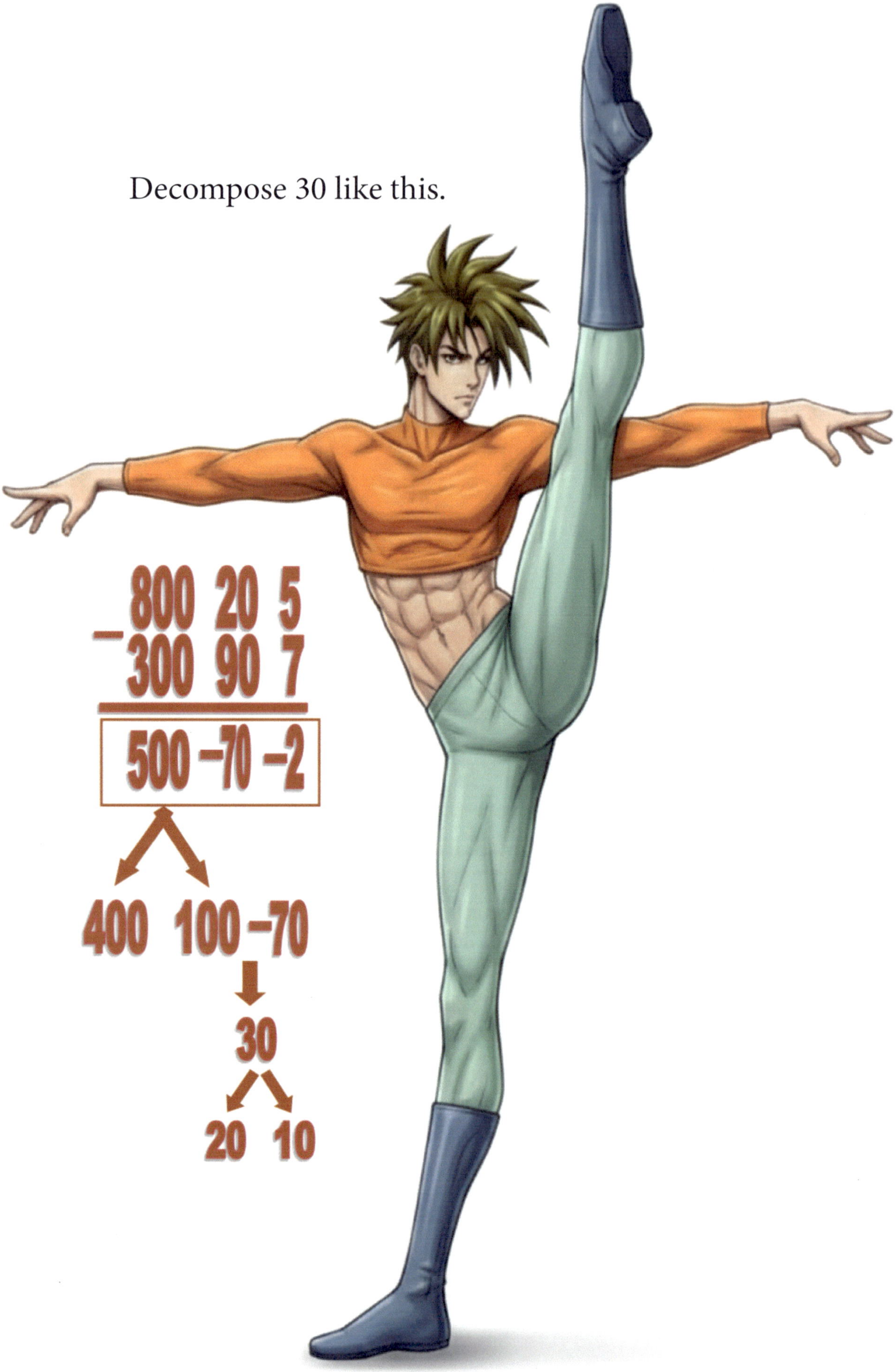

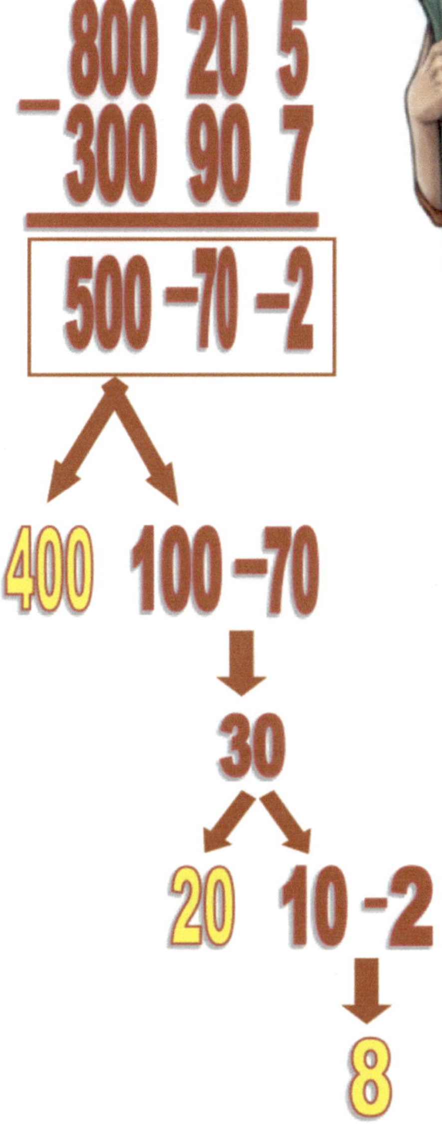

Subtract 2 from 10.

Write its difference.

Therefore, the answer to 825—397 is equal to 428.

$$\begin{array}{r} 825 \\ -\,397 \\ \hline \boxed{428} \end{array}$$

Excellent!

Now, let us discuss the
strategy.

What is the Compensation and Place Value Decomposition Technique?

Let me explain.

The Compensation and Place Value Decomposition is a subtraction method that works that is use to avoid traditional borrowing (regrouping), making the mental math much simpler.

The goal shown is to solve 95 - 28. Here is the breakdown of why those steps lead to the correct answer of 67:

1. Compensation (Rounding Up)

The method starts by turning the "difficult" number (28) into a "friendly" number (30).

To get from 28 to 30, you add 2. In subtraction, if you subtract a larger number than intended, you have to add that difference back at the end to stay balanced.

2. Decomposing the Numbers

The image breaks the problem into tens and ones:

Tens: 90 - 30 = 60

Ones: The original 5 from 95, plus the 2 we used to round up the 28.

This creates two separate, easy problems: 90 - 30 and $5 + 2.

3. Combining the Results

Finally, you add the results of those two groups together:

60 + 7 = 67

Why does this strategy work?

This method works because it uses Constant Difference rule or Simple Associative Properties.

The Logic

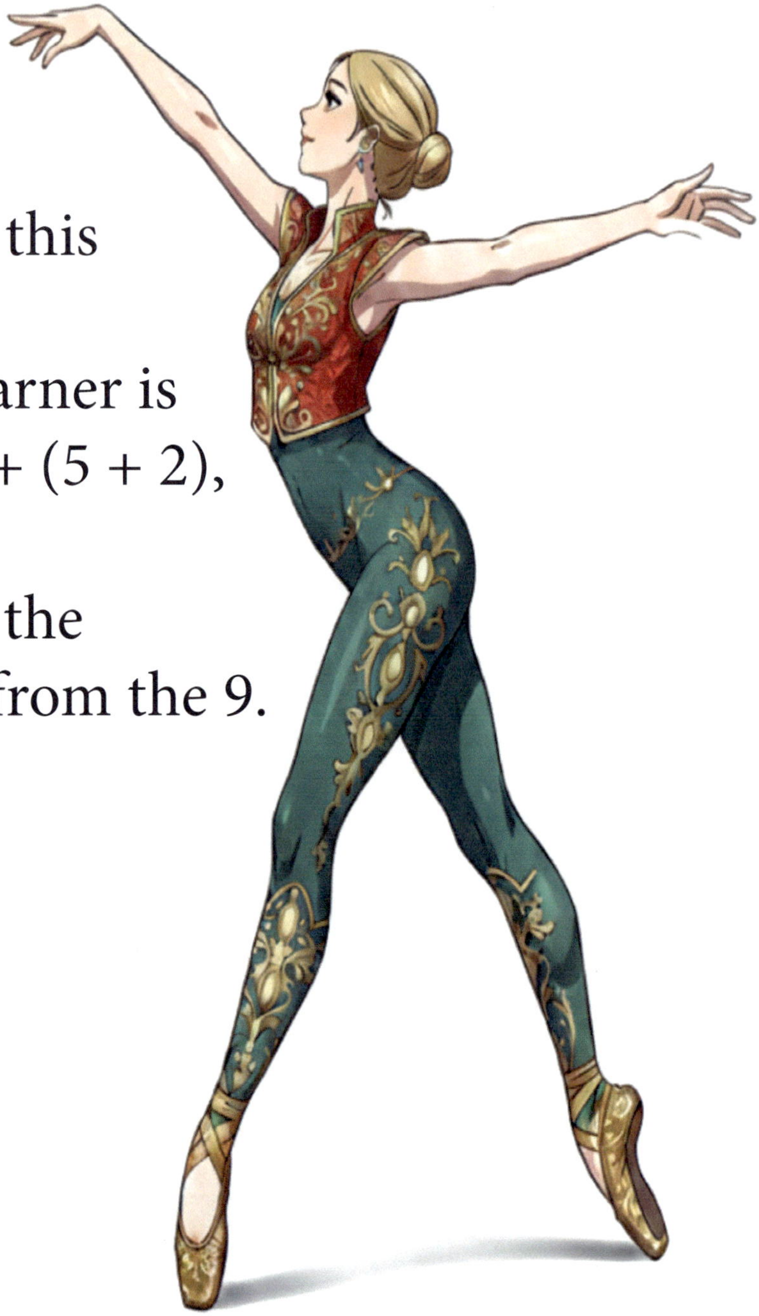

We can see the logic of this method by splitting it into 90 and 5. Doing this the learner is essentially doing (90 - 30) + (5 + 2), which arrives at the same logical conclusion without the messiness of "borrowing" from the 9.

What is the advantage of this technique over traditional regrouping or borrowing technique?

The technique is a form of mental math subtraction that offers several cognitive and practical advantages over the traditional "borrowing" (regrouping) method.

Here are the primary advantages:

1. Reduces Cognitive Load.

Traditional regrouping requires you to hold multiple "invisible" numbers in your head (e.g., crossing out a 9, writing an 8, and carrying a 1 to make 15).

The Advantage: This technique simplifies the problem into "friendly" numbers ($90 - 30$). By rounding the number being subtracted, you eliminate the need to track small remainder numbers, which reduces the chance of losing your place mid-calculation.

2. Prevents "Place Value" Errors

In traditional borrowing, students often get confused about which column a number belongs to, leading to mistakes like subtracting the smaller top digit from the larger bottom digit (e.g., seeing 95 - 28 and thinking 8 - 5 = 3).

The Advantage:
By decomposing the numbers into their full values (90 and 30), it reinforces number sense. You are dealing with the actual magnitude of the numbers rather than just isolated digits in columns.

3. Faster for Mental Math

The traditional method is designed for paper and pencil; it is very difficult to "see" a crossed-out 9 and a carried 1 in your mind's eye. The Advantage: This method moves from left to right (subtracting the tens first), which is how our brains naturally process magnitude. It allows you to calculate 95 - 28 as 65 plus 2 equals 67 almost instantly.

4. High Accuracy with "Friendly Numbers"

Subtracting multiples of 10 is significantly easier and less prone to basic arithmetic errors than subtracting mixed digits like 8 from 15.

The Advantage: You are essentially trading one "hard" subtraction problem (15 - 8) for one "easy" subtraction (90 - 30) and one "easy" addition (5 + 2).

Comparison at a Glance

FEATURE	TRADITIONAL REGROUPING	ACON's WAY
DIRECTION	RIGHT TO LEFT (ONES THEN TENS)	LEFT TO RIGHT (TENS THEN ONES)
MENTAL EFFORT	HIGH (MUST REMEMBER BORROWED DIGITS)	LOW (USES "FRIENDLY" NUMBERS
TOOLS NEEDED	USUALLY REQUIRES PAPER AND PENCIL	OPTIMISE FOR MENTAL CALCULATIONS
CONCEPT	PROCEDURAL (FOLLOWING STEPS)	CONCEPTUAL (UNDERSTANDING NUMBER VALUES)

What does research say
about this technique?

Research on the compensation technique highlights its effectiveness in building mathematical flexibility and number sense, particularly when compared to standard algorithms.

1. Superior Efficiency and Accuracy

Studies indicate that students who use the compensation strategy often perform calculations faster and just as accurately as those using other mental methods like the "jump" strategy.

Target Numbers:

Research shows this method is particularly efficient for problems where the number being subtracted (subtrahend) ends in an 8 or 9 (e.g., subtracting 28 or 19), as these are close to "friendly" multiples of 10.

Reduced Errors:

Traditional regrouping often leads to errors where students simply subtract the smaller digit from the larger one regardless of position (e.g., 95 - 28 = 73); compensation avoids this by focusing on the total magnitude of the numbers.

2. Cognitive and Developmental Benefits

Educational research emphasizes that these number-based" strategies differ fundamentally from "digit-based" traditional algorithms.

Number Sense:

Unlike the traditional algorithm, which can encourage "cognitive passivity" or just following ‘ steps, compensation requires students to actively reason about relationships between numbers.

Mathematical Reasoning:

There is a significant positive correlation between a student's ability to perform mental computations like compensation and their overall mathematical reasoning skills.

Flexibility:

Research suggests that teaching a variety of strategies (like compensation and indirect addition) helps students become "adaptive" problem solvers who can choose the easiest method for a specific task.

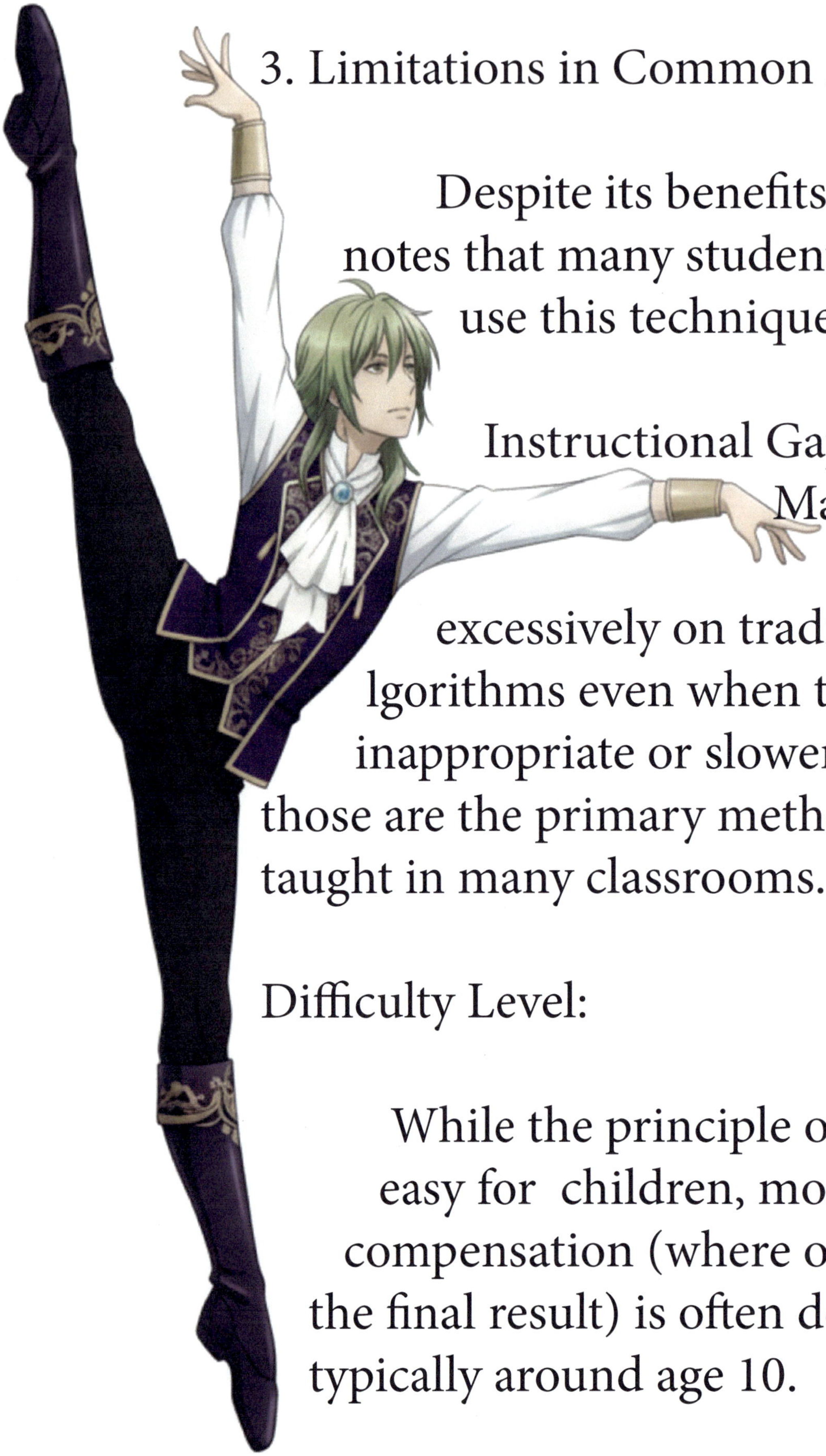

3. Limitations in Common Practice

Despite its benefits, research also notes that many students struggle to use this technique spontaneously.

Instructional Gap: Many students rely excessively on traditional lgorithms even when they are inappropriate or slower because those are the primary methods taught in many classrooms.

Difficulty Level:

While the principle of "Identity" is easy for children, more complex compensation (where one must adjust the final result) is often developed later, typically around age 10.

What is Decomposition and Partial Negative Difference technique?

The Decomposition and Partial Differences Technique is a strategy that is use to break a complex subtraction problem into smaller, manageable chunks. This specific approach is a form of mental math that bypasses traditional "borrowing" by allowing for negative partial results.

1. Place Value Decomposition

The method begins by breaking both numbers into their tens and ones:

$95 becomes 90 and 5.

$28 becomes 20 and 8.

2. Calculating Partial Differences

Instead of regrouping, you subtract each column independently:

Tens: 90 - 20 = 70.

Ones: 5 - 8 = -3.

This is the "secret sauce" of the method; it acknowledges that you are "short" by 3 in the ones column rather than trying to force a borrow from the 90 immediately.

3. Recomposing the Total

To find the final answer, you simply combine your partial results:

70 + (-3) is the same as 70 - 3.

In the example, 70 is split into 60 and 10. Then, 0 - 3 = 7.

Finally, 60 + 7 =67.

Why It's Effective

1. It is effective because it maintains Place Value: You are always working with the actual value of the numbers (90 and 20) rather than just digits (9 and 2).

2, Left-to-Right Processing: Like the compensation method, this allows you to calculate the "big" part of the number first, which is more natural for mental estimation.

3. No "Invisible" Numbers: You don't have to remember a "1" carried over or a "9" turned into an "8"—every step is written out based on the actual values.

How about the Partial Difference
of Partial Subtraction?

What does the research
say about it?

Research into this specific technique—often called the Partial Differences or Partial Subtraction method—shows that it provides a strong foundation for mental flexibility and number sense.

1. Accuracy and Efficiency

Perfomance: Studies indicate that students using these types of mental strategies often calculate more quickly and maintain a level of accuracy comparable to those using traditional methods.

Effective Scenarios: This method is particularly effective when the subtrahend (the number being subtracted) ends in a high digit like 8 or 9, where traditional borrowing would be required.

Error Prevention: It helps prevent "smaller-from-larger" errors common in traditional regrouping, where a student might mistakenly calculate 95 - 28 as 73 by subtracting 8 - 5.

2. Cognitive and Conceptual Development

Place Value Awareness: Research highlights that "number-based" strategies like this one force students to work with the actual value of the numbers (e.g., 90 and 20) rather than just isolated digits (9 and 2), which significantly improves number sense.

Active Reasoning:

Unlike traditional algorithms that can lead to "cognitive passivity," this technique requires the student to actively reason about the relationships between the numbers.

Predictor of Success: A student's ability to use mental strategies like this is a strong predictor of their overall mathematical reasoning skills.

3. Instructional Challenges

Lack of Spontaneity: Despite the benefits, research shows many students do not use these methods spontaneously because they are often over-trained on traditional algorithms in the classroom.

Developmental Timing: While the basic principles are easy to grasp, the specific skill of adjusting the final result (compensation) usually develops more fully around age 10.

Strategic Flexibility: Educational experts emphasize that teaching these varied strategies allows students to become "adaptive," choosing the most efficient path for any given problem.

How do you classify the
strategy used on pages 42 to 53?

The strategy shown on pages 42 to 48 is known as the Constant Difference or Equal Addition method.

Why Does This Strategy Work?

This stragy work because it preserves the exact distance between two number on a number line by adjusting both numbers by the same amount.

Why It Works Mathematically?

It works mathematically because it observes the fundamental principle which state that the difference between two numbers, a and b, remains unchanged if you add the same value (k) to both.

Mathematically, it looks like this:

$$(a + k) - (b + k) = a - b$$

In example of 864 - 478:

First Adjustment:

Add 2 to both numbers to turn 478 into a "friendly" number (480). Now the problem is 866 - 480.

Second Adjustment:

Add 20 to both numbers to reach the next "friendly" hundred (500). The problem becomes 886 - 500.

The Final Result:

Since subtracting 500 from 886 is a simple mental task, you quickly arrive at 386.

Scientific and Educational Advantages

Research highlights several reasons why this strategy is often superior to traditional borrowing for mental and conceptual math:

Elimination of Borrowing Errors:

Traditional regrouping often leads to "smaller-from-larger" errors (e.g., a student might see 4 - 8 and just write 4). This method avoids that by shifting the numbers away from difficult "regrouping" zones.

Efficiency with High Subtrahends:

Studies show this is the most efficient strategy when the number being subtracted ends in an 8 or 9, as it requires the fewest steps to reach a round number.

Reduced Cognitive Load:

Unlike borrowing, which requires holding "carried" digits in your head, this method focuses on total magnitude, making it much easier to perform mentally.

Development of Number Sense:

It requires students to actively reason about how numbers relate to one another rather than just following a memorized procedure.

How do you classify the strategy used on pages 49 to 53?

The solution on pages 49 –53 is a subtraction strategy calle Decomposition and Partial Differences. This is used by breaking a complex multi-digit subtraction problem into smaller, simpler parts that can be solved mentally.

How the Strategy Breaks Down Using the example of 864 - 478, the process works by subtracting one place value at a time:

Subtract the Hundreds:

864 - 400 = 464.

This leaves you with 464 and 78 still to subtract. Subtract the Ten (with Compensation):

To make subtracting 78 easier, 78 is rounded to 80. To keep the math balanced, 464 is also increased by 2 to become 466.

466 - 80 = 386.

The Result:

By the final step, the problem has been simplified to 486 - 100, arriving at the correct answer of 386.

Why Research Supports This Method?

This "number-based" approach is often more effective than traditional borrowing for several reasons:

Maintains Number Sense: Students work with the full value of the numbers (like 400 or 80) rather than just isolated digits (4 or 8), which deepens their understanding of place value.

Reduces Cognitive Load: It avoids the need to keep track of "regrouped" or "carried" numbers in your head, which is where many mistakes happen in traditional borrowing.

Encourages Flexibility: Studies show that students who use these "invented" strategies are often more accurate and develop better mathematical reasoning skills than those who only memorize the standard algorithm.

Faster Mental Processing: This method follows a left-to-right logic (subtracting the largest values first), which is how the brain naturally estimates and processes magnitude.

How about pages 64 –69?

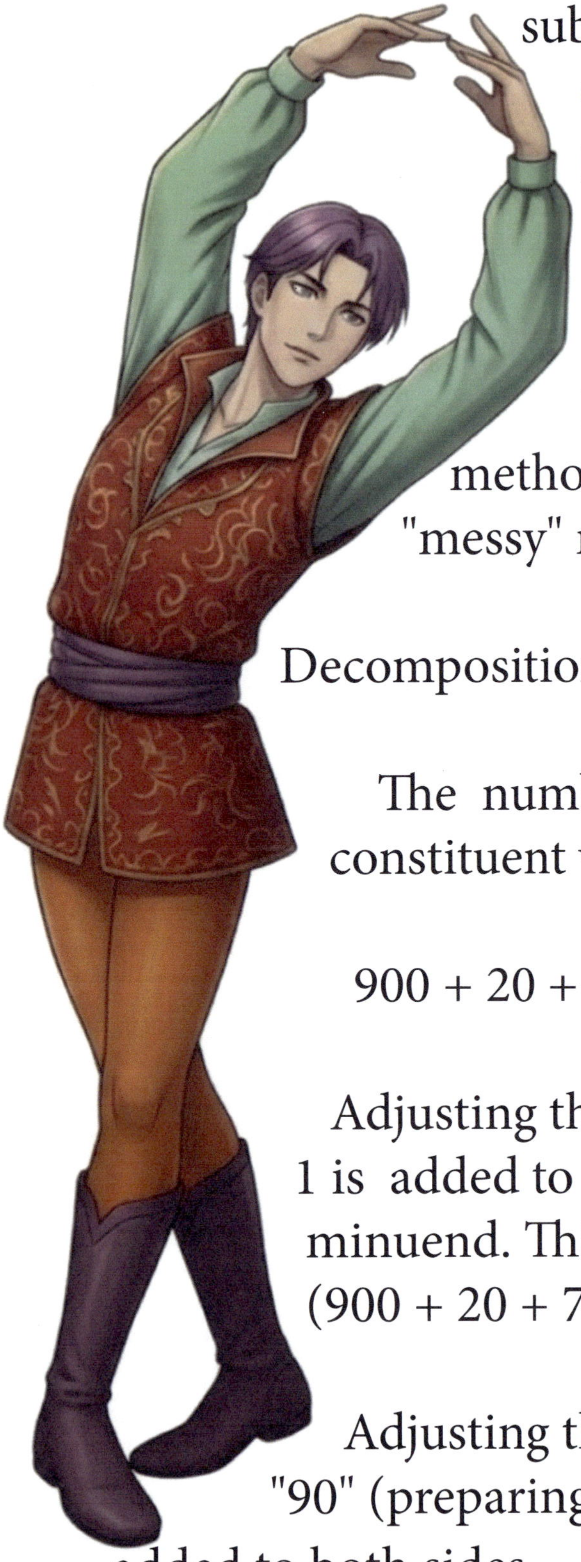

The strategy on pages 64-69 uses multiple rounds of compensation to transform a difficult three-digit subtraction problem into one that can be solved instantly through place value.

Why It Works?

In the problem 926 - 489, the method follows a logical path to eliminate "messy" numbers.

Decomposition:

The numbers are first broken into their constituent values:

900 + 20 + 6 and 400 + 80 + 9.

Adjusting the Ones: To get rid of the "9" in 489, 1 is added to both the subtrahend and the minuend. This changes the problem to (900 + 20 + 7) - (400 + 80).

Adjusting the Tens: To turn the "80" into a "90" (preparing for the next round), another 10 is added to both sides.

Final Rounding: Eventually, the bottom number is rounded up to the nearest hundred (500). To maintain the constant difference, the top number is also increased.

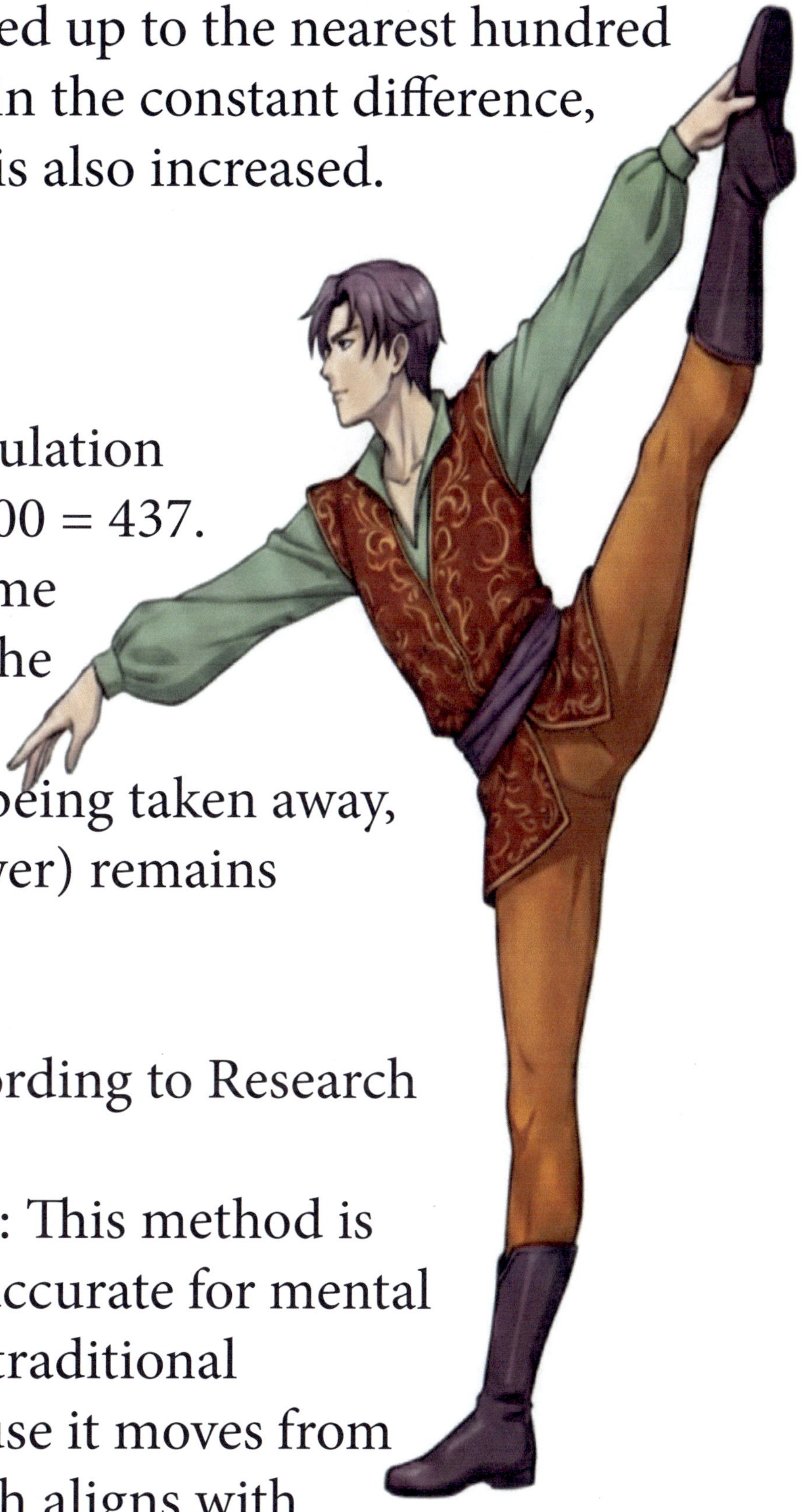

The Result:

The final calculation becomes 937 - 500 = 437. By adding the same amount to both the starting total and the amount being taken away, the gap (the answer) remains identical.

Benefits According to Research

Mental Accuracy: This method is faster and more accurate for mental calculation than traditional regrouping because it moves from left to right, which aligns with how our brains process magnitude.

Number Sense:

Research highlights that strategies like this prevent students from making "digit-based" errors (like subtracting 4 from 9) because they are constantly interacting with the actual value of the numbers.

Cognitive Load:

By converting the subtrahend into a "friendly" number like 500, the student significantly reduces the mental energy required to finish the problem.

How about pages 74 –80?

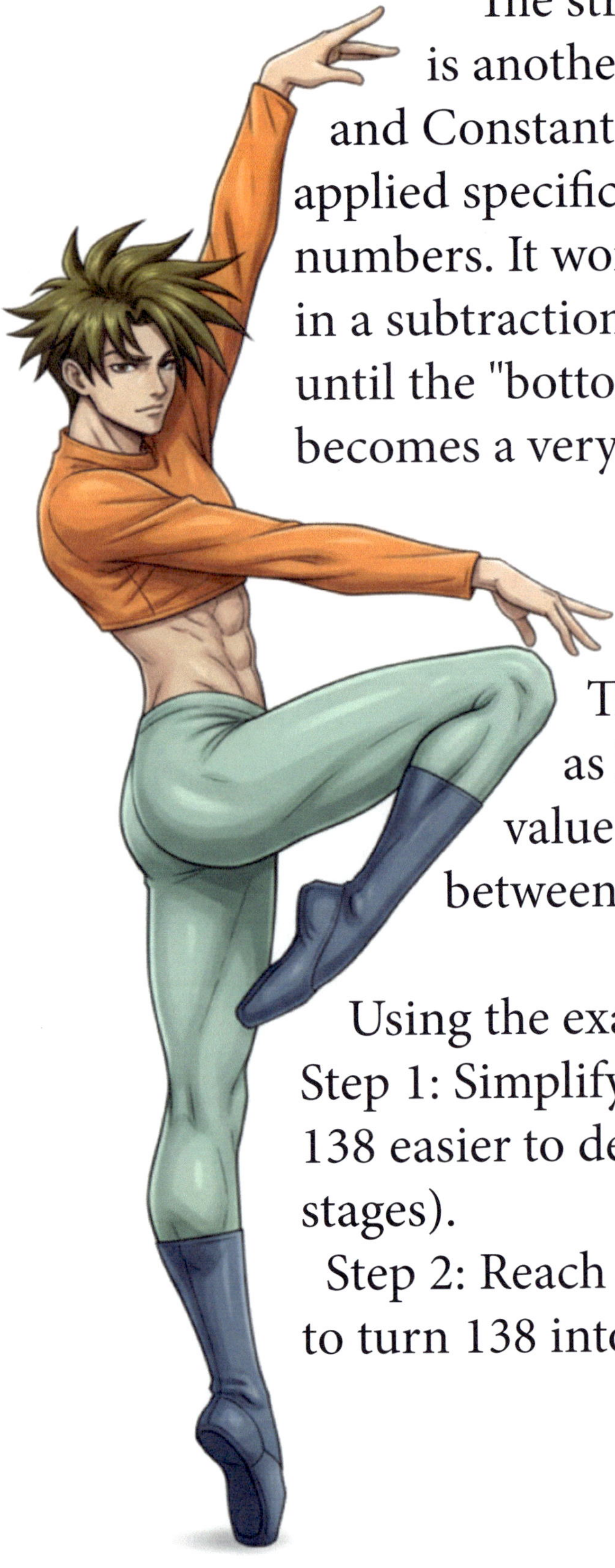

The strategy used on pages 74 –80 is another variation of Compensation and Constant Difference, but it is applied specifically to three-digit numbers. It works by adjusting both numbers in a subtraction problem by the same amount until the "bottom" number (the subtrahend) becomes a very simple multiple of 100.

Why It Works Mathematically?

The core logic remains the same: as long as you add the same value to both numbers, the difference between them remains identical.

Using the example on page 74: (425 - 138):
Step 1: Simplify the Tens: To make the 30 in 138 easier to deal with, 62 to both sides (in stages).
Step 2: Reach a Friendly Hundred: The goal is to turn 138 into $200.

Result: Subtracting 200 is a basic place-value task that results in 287.

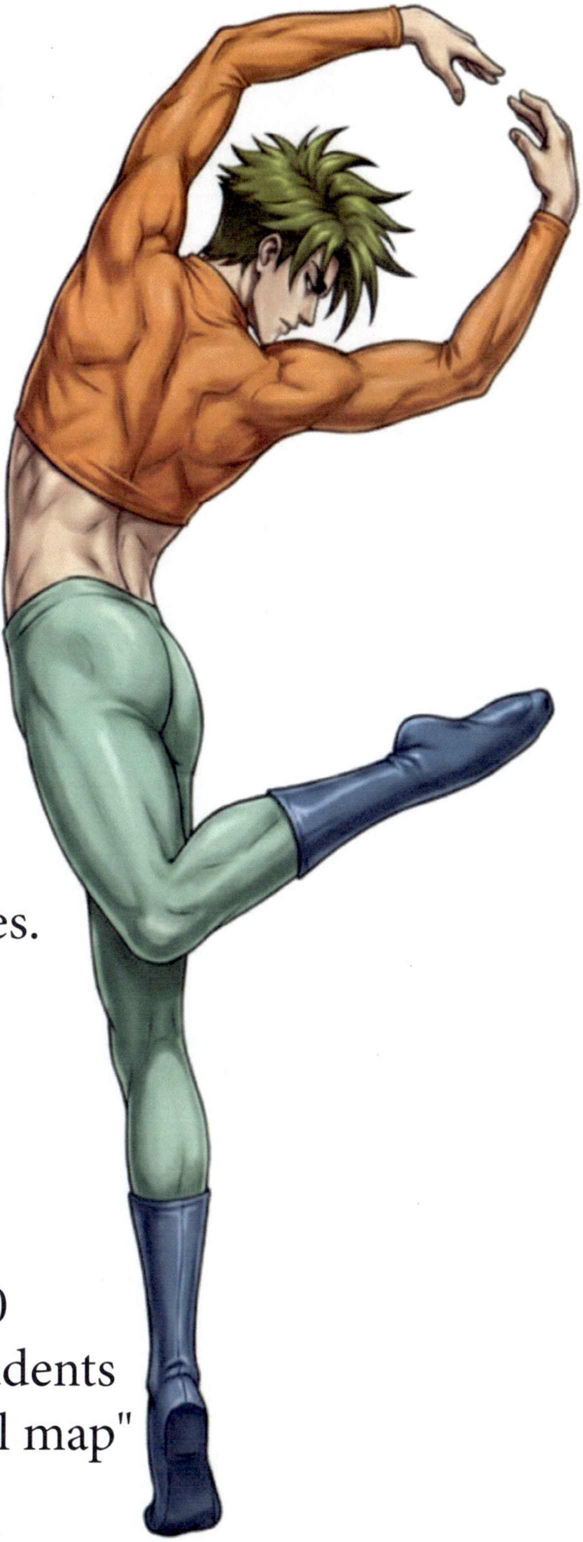

Why Research Supports this "Big Picture" Approach?

Educational research highlights that this specific three-digit approach is powerful for several reasons:

Reduces "Borrowing" Errors: Traditional regrouping with three-digit numbers (like 425 - 138) often requires "double borrowing," which is a high-risk area for student mistakes. This method eliminates that risk entirely.

Focuses on Magnitude:

By working with 400 and 200 rather than just digits 4 and 1, students develop a much stronger "internal map" of how large numbers behave.

Mental Scalability:

Studies show that once a student masters this for two-digit numbers, they can easily scale it to three or four digits, whereas traditional borrowing becomes exponentially more difficult for the brain to track mentally as digits are added.

Cognitive Efficiency:

Turning the subtrahend into a "friendly" hundred significantly lowers the mental effort required to finish the problem.

***END

Made in the USA
Las Vegas, NV
03 November 2026

57422972R00098